Hall 3

Christopher L Hall — SPRING 94

DESCRIPTIVE TITLE OF COURSE	DEPT.	NUMBER	MARK	SEM. HRS.
Chem Colloq II	CH502	001	A	.50
Reac Org Compd	CH538	001	A	3.00
Adv Phys Chem	CH546	001	A	3.00
Spec Topics II	CH582	001	A	3.00

HRS. ATTEMPTED: 9.50

	SEM	CUM	SEM	CUM	SEM	CUM
CREDITS	9.50	9.50				
QUALITY POINTS	38.00	38.00				
G.P. AVERAGE	4.00	4.00				

Christopher L Hall — FALL 95

DESCRIPTIVE TITLE OF COURSE	DEPT.	NUMBER	MARK	SEM. HRS.
Colloquium	CHEM 5			
Diff Equations	MATH 271	001	A	
Intro Mod. Opti	PHYS 472	001	A	

HRS. ATTEMPTED: 7.50

	SEM	CUM
CREDITS	7.50	40.00
QUALITY POINTS	30.00	148.00
G.P. AVERAGE	4.00	4.00

Christopher L Hall — FALL 94

DESCRIPTIVE TITLE OF COURSE	DEPT.	NUMBER	MARK	SEM. HRS.
Chem Colloq I	CH501	001	A	.50
Adv Organic	CH535	001	A	3.00
Adv Physical	CH545	001	A	3.00
Mol Spect	CH547	001	A	3.00
Chem Rsrch I	CH591	001	A	5.00

HRS. ATTEMPTED: 14.50

	SEM	CUM
CREDITS	14.50	24.00
QUALITY POINTS	58.00	96.00
G.P. AVERAGE	4.00	4.00

Christopher L Hall — SPRING 96

DESCRIPTIVE TITLE OF COURSE	DEPT.	NUMBER	MARK
Thesis Prep	CHEM 599	001	P

HRS. ATTEMPTED: .00

	SEM	CUM
CREDITS	.00	40.00
QUALITY POINTS	.00	148.00
G.P. AVERAGE		

Christopher L Hall — SPRING 95

DESCRIPTIVE TITLE OF COURSE	DEPT.	NUMBER	MARK	SEM. HRS.
Chem Colloq II	CH502	001	A	.50
Chem Rsrch	CH592	001	A	5.00
Int Elect/Magn	PH352	001	F	3.00

HRS. ATTEMPTED: 8.50

	SEM	CUM
CREDITS	8.50	32.50
QUALITY POINTS	22.00	118.00
G.P. AVERAGE	4.00	4.00

GRANTED HONORABLE DISMISSAL _____ WITHDREW _____ REASON _____

MAJOR _____ DEGREE _____ DATE _____ RANK _____ IN CLASS OF _____

HONORS _____

THE ABOVE IS A TRUE AND ACCURATE RECORD. IT BECOMES OFFICIAL ONLY PROPERLY SEALED AND SIGNED.

PERMANENT RECORD

a ward of the court

Dr. Christopher Hall, MD

Contents

Introduction	5
Chapter 1: Childhood	8
Chapter 2: A Letter from Royce Wayne Timmons	38
Chapter 3: College	53
Chapter 4: Dental School	76
Chapter 5: Medical School	100
Chapter 6: Residency	116

Introduction

I first thought about writing a book in my early twenties. I was a dental student at Northwestern University in Chicago and the thought came to me—"Wow! This has been an interesting journey!" There I was, sitting in the college cafeteria with snowflakes blowing about 15 feet in the air past the window. Having been born and raised in California, I was far from sunny Los Angeles. It was the middle of fall and we had just completed our gross anatomy exam. At 23 years old, I had lived probably more than the average person my age. At four, I was declared a ward of the court. My sisters, brothers, and I had been found left without parental supervision at home by the department of social services, our mother gone. I really don't remember much about my life before four years old. I remember being barefoot and being taken by my social worker to my very first foster home. That was the beginning of a journey that would include two more foster homes, four boys' homes, and multiple stints in three different Juvenile Hall facilities.

I was a rebellious kid, I must say. Basically, what I mean is that I really did not want to be adopted by anyone. I'm sure if I would have simply gone along with the Department of Social Services' intent, I would have likely grown up in one foster home and that would have been it. Sometimes I wish I would have been wise enough at four to realize that staying with one foster family would have made my life much easier.

Well, you live and learn. Consequently, I never lived in one city more than three years, and I ended up attending three different high schools. Those were painful years of my life. I've obviously moved on and made something of myself; however, the journey has not been easy. My paternal great aunt once told me "You are going to have a hard time." She was looking at a twelve-year-old boy at the time with no parental support and with no one to help him plot the course of his life. It really was not until the age of fourteen that I realized I was becoming a man. I had to start thinking about how I would support myself once the state of California released me from the Department of Social Services. My first inclination was to become a lawyer. I must have gotten this idea at around eight years old. I used to watch *Perry Mason* and *Dragnet*, and they always portrayed lawyers as people who wore suits. Someone wearing a suit looked like a very important person to me at eight years old. Years later in college, I realized that my real interest was in science and not law.

Well, what follows is my story and I hope that it is an inspiration to someone. I hope that at least one young person sitting out there in foster care, the juvenile justice system, or a boys' or girls' home will be inspired.

Chapter 1: Childhood

I was born in Los Angeles, California at the General Hospital. My father's name was Luther Henry Hall, and my mother's name is Bessie Marie Hall. I'm not sure what year my parents were married, nor for how long. My parents are originally from a small town called Houma in Louisiana. I've only been there once and that was when I was sixteen years old.

It is my understanding, as my older sister Joyce tells me, that my father originally came to California in the late 1950s looking for better opportunities for our family. I once saw a letter written by my father explaining how he had found land in California and wanted to move the whole family to Los Angeles. My father had apparently finished high school but was not college educated. On my birth certificate he is listed as a janitor.

My mother's background is very similar. As I understand it, she finished high school once we came to

California; she had gotten pregnant at the age of fourteen. She never attended college. My older brother Wayne and my sister Joyce were apparently born in Louisiana.

The Sixties were obviously a very turbulent time. Sometime during the first four years of my life, my father was imprisoned for killing a man. My understanding is that this occurred over a card game. My dad apparently hit another man over his head with a club and killed him. He was sent to prison for manslaughter and my brothers, sisters, and I were left with our mother. Unfortunately, my mother was like so many poor, single mothers today. She was uneducated and unemployed. At the age of four I was declared a ward of the court.

I remember my first foster mother. Her name was Mrs. Hankins and she lived on Somerset Drive in Los Angeles. She was married to Virgil Hankins and they had no children of their own. Mrs. Hankins was a rather obese woman. Mr. Hankins, on the other hand, was a thin man. So, at the age of four, barefoot and in the rain, I was introduced to Mrs. Hankins. I asked her if I could call her mother and she said that was fine with her. She had a small house; there were two bedrooms and one bathroom. She was very strict, but she would take me on outings to the Crenshaw Shopping Center. Her husband, Virgil, was a gardener. He was very short-tempered and had little patience with children. I would go with him on his different landscaping assignments and he would teach me how to cut hedges and sweep up grass. I learned a lot of gardening from him.

Overall, my stay with Mr. and Mrs. Hankins was very nice. We had some neighbors next door, and their children were also foster children. I particularly remember an Asian boy named John. I remember him because he and I had gotten into a rock fight and he hit me on top of my head with a sharp rock. I still have a scar from that incident. I lived with Mr. and Mrs. Hankins from about the age of four to the age of ten.

I did run away from the Hankins home on a number of occasions. Mrs. Hankins was a good foster mother, but I was a child who was restless and bored easily. I enjoyed running away and exploring the world on my own. It gave me a sense of freedom, and I guess living with a foster mother, a person who was not my birth mother, made me feel unwanted. Therefore, I would run away.

I was very young when I began running away from foster homes. Maybe I felt that I would eventually find my original family if I ran away enough. It is hard to explain the feelings that one has as a foster child. I know at times I felt lost, confused, and hopeless. Why, I wondered, was I in a foster home? There were many questions that I would not learn answers to for at least another ten years.

While living with Mr. and Mrs. Hankins, I attended Virginia Road Elementary School. My favorite teacher was Mrs. Gillman. I remember her being sweet and speaking very softly to me. She was my third grade teacher. In fourth grade, I had a teacher named Mrs. Coleman who was very strict. She would hit us on our knuckles when we acted up. Occasionally, I think about my childhood friends from that time. I remember a girl named Michelle Bowden. She was my first crush, and I think this occurred in the fourth grade. Michelle Bowden was always dressed nicely in plaid dresses and appeared very mature for her age.

    I ran away from the Hankins home on one occasion and went to Michelle's house. I didn't go in and I don't think she knew I was even there. I was across the street looking at her on her front porch. The Bowdens had a large two-story home and I wondered what it was like to live with brothers and sisters in a nice house.

    I think eventually Mr. and Mrs. Hankins became tired of me running away and eventually requested that the Department of Social Services remove me. So, at the age of ten years old I was moved to another foster home with the Allen family. Years later, when I was in high school, I would occasionally stop by and visit Mrs. Hankins. I was naïve thinking she might be happy to see me. My social worker at that time contacted me and told me that Mrs. Hankins did not want me to stop by and see her anymore.

The Allens' home was much larger than the Hankins' home. The Allens lived about three miles away from the Hankins. There were more kids in the Allen home, six of us in all—three boys and three girls, including myself. The Allens were younger than Mr. and Mrs. Hankins; I believe they were in their forties. The home was also very strict. Mr. Allen would not put up with too much acting out from his children. We would get beaten with a belt regularly. Usually we all received a beating at once so that you could hear the person before you crying. Mr. Allen also did not believe in breaking up fights. When the boys would get in a fight he let it continue until one would give up or was beaten up badly enough. I do remember the Allens taking us to the beach and shopping on many occasions.

I first went to Venice Beach in California while living with the Allens. If you've ever been to Venice

Beach, you probably understand why as a child this became one of my favorite places to hang out. There were magicians, palm readers, vendors, and acrobats along the beach. There was a very long, smooth concrete pathway that extended from Venice Beach to Santa Monica and beyond for skating. Venice Beach, to me, at that age, represented endless freedom. Well, this was a favorite place for the Allens to visit.

      I enjoyed my time with the Allens; unfortunately, I don't remember the name of the school I attended during that time. My routine of running away continued and I must have run away more than a dozen times while I lived with the Allens. I guess you may be wondering what I was doing when I ran away—mainly stealing from stores. On several occasions, another boy and I would run away together and we would visit grocery stores to steal candy and toys. We would wear large clothing and stuff as much candy, toy cars, machine guns, and other toys into our jackets as we could. Most of the time, we would get away; however, we would occasionally get caught by the store security.

Throughout my stay with the Allens, my older sister Joyce would come and pick me up during the holidays. During those times, I would have the time to bond with my brothers and sisters and experience family interactions. I never really knew my mother very well growing up. She always seemed a distant figure to me. My mother had a stroke in her twenties. She is paralyzed on the right side of her body. Why my mother had a stroke at such a young age, I am not sure. My sister Joyce believed it was from my father knocking her over her head one too many times.

My real father was a very abusive man. He apparently beat my mother often. Joyce told me she used to get beatings also since she was not one of my father's children. Joyce and my brother Wayne had a different father from my sister Sandra and I—we have the same mother and father. I will talk a little more about my father later.

My foster parents, Mr. and Mrs. Allen, apparently did not want me to visit with my real family. I remember one time Joyce and her common law husband James came to pick me up for a weekend and the Allens hid me in the back of the house and told my sister that I was not home. The fact is that I really looked forward to seeing my real family. Living in a foster home away from my real family always kept me wondering about why and how I had ended up in this situation, separated from my family.

I ran away from the Allens on many occasions. Sometimes I would stay gone a week or more. I would sleep in abandoned cars or sometimes I would sleep in a cardboard box. You simply had to fold yourself up in a tight ball and sit in the box. I eventually came to the end of my rope with the Allens. While I'm sure they were tired of me running away, I was forced to leave their home for another reason. One of my foster brothers named Tony had burned their garage down and, unfortunately, I was with him when he did it. My recollection of the incident is as follows: He and I were in the garage. He was somewhere in the back behind some mattresses and I could not see him. He ran out of the garage and said "Run, Chris! It's a fire!" I ran out of the garage only to find the whole structure on fire. We ran in to tell our foster parents, who promptly called the fire department. The garage was burned to the ground before the firemen arrived. The firemen determined that arson was the cause. I still believe to this day that Tony set that fire. However, we were both blamed for it and, therefore, off to Juvenile Hall we went.

    Juvenile Hall is not the place a young person wants to end up. It is a very lonely place, particularly in a city like Los Angeles with its rivalries between Hispanic and Black gangs. Juvenile Hall is in fact a prison for youth offenders. The walls are uniform and at that time in the 1970s were painted similar colors. All inmate wore the same color uniform and were marched in a single file line to the mess hall and back.

Each inmate had a small room which had a toilet and a sink and one bed with an extremely thin mattress. The door to each room had an eight by eleven peep hole covered with glass-plated wire. At nighttime, a very dim light would be left on in each room so that guards could see the inmate during their nightly checks.

In Juvenile Hall, there is a lot of fighting. Rival gang members encountered each other there. Sometimes a fight would break out just because two boys looked at each other wrong.

So, my first stay in Juvenile Hall was about six months long. I didn't interact with many of the boys there. I did a lot of reading in my room after lockdown and played a lot of ping-pong during the day. The food in Juvenile Hall was terrible; I particularly remember the corn being soggy and sticky. I don't remember having phone privileges, and I am not sure they even granted them. The first book I ever read cover to cover was when I was in Juvenile Hall. The book was called *Where You Stand is Where You Live*. It followed the life of a runaway from the United States all throughout England. I saw myself in that character—a runaway wandering, searching, and seeking, with no end in sight.

I was released from Juvenile Hall when I was about eleven years old and placed in another foster home. I don't remember the names of these foster parents, and I believe I only stayed in this home for about a week. I remember jumping out of the bedroom window to escape and disappearing into the night. I was eventually caught and sent to another home where I also promptly ran way.

Soon, the Department of Social Services thought it best if I were placed in a children's psychiatric unit at Riverside Hospital. I am not sure why I ended up at that place. Nevertheless, I made the best of my stay there. I do remember some of the residents there, though. One resident who I interacted with a lot was named Lupe. She was a Hispanic girl who had experienced a tough childhood and who was into self-mutilation. She would often cry, and she and I became good friends. I was only eleven years old and small in stature, and she used to set me on her lap and treat me like a little brother. One of the counselors (I believe his name was Bill) also took me under his wing. He treated me like a son, and I remember that years later when I was in a boys' home, he came and visited me and brought me a bicycle. Overall, my stay at Riverside Hospital was short and I remember being sad to have to leave there.

    I was moved from Riverside Hospital to Children's Baptist Home in Inglewood, California. Children's Baptist Home was a huge complex. There were five cottages and one group home located at the facility. It housed boys from ages eight to fourteen. The administrative building was huge and it had several offices for the facility president and about ten or twelve social workers. Each cottage had about six staff that would work eight hour shifts. In the middle of the complex there was a huge swimming pool. The facility had a huge baseball diamond in the back and a massive dining facility. There were even classrooms in the facility for children to attend school. Many boys who live in state facilities such as Children's Baptist Home have behavioral problems that prevent them from attending public school. Sometimes these children are simply academically too slow to keep up in a normal public school.

My social worker at Children's Baptist Home was named Harold. He was a young white man who lived on Venice Beach. He had a full head of hair styled in an Afro. I was introduced to him on my first day at Children's Baptist Home, and he escorted me to my new home, cottage three.

Looking back over the years, I realize now that it was my experiences at Children's Baptist Home that made me realize I could be almost anything I wanted to be in this life. Why do I say this? Well, Children's Baptist Home had many intramural sports that residents could participate in. Of course, then my dream, like most young boys, was to grow up and be a football or baseball star. Becoming a lawyer was a distant third. I do believe, though, that it was the abundance of these activities that made me realize that I could do almost anything. I participated in football, basketball, swimming, scouting, dirt bike riding, and ping-pong. My first baseball mitt I ever received was at Children's Baptist Home.

    My days at Children's Baptist Home were fairly routine. A typical day would start about 7:00 AM. We would awaken, wash up, and get dressed. Then we would walk in a single file line down to the cafeteria, which was in the center of the complex. Each cottage would sit at their individual table. Once we finished breakfast, we would return to the cottage and then head off to class. The mess hall in almost every boys' home is a place for potential problems and Children's Baptist Home was no exception. The mess hall was a place where fights frequently broke out. In the mess hall, boys from each cottage could mix, and I think other boys knew fighting in the mess hall would provide a stage to show other boys who was boss. I will talk a little more about this later on.

After breakfast, we would proceed to the classrooms in the facility. I initially began going to school at the facility and after about six months they transferred me to public school. I was clearly bored in those in-facility classrooms. This was a pattern that would continue from one boys' home to another—I would eventually be allowed to attend public school. After school, I would return to Children's Baptist Home and participate in any number of activities. I was very active in scouting and was eventually made senior patrol leader. Becoming a Boy Scout was another turning point in life for me. Julio was our scout master.

Julio was young, about twenty-seven. He took us on many camping trips. I was the senior patrol leader of Troop 199 in Inglewood, California. It was my responsibility to conduct all drills, regulate and plan troops activities, and make sure while at camp sites that the sites were left clean. Our troop had four patrols and each patrol had a patrol leader who had to report to me. Scouting helped me develop leadership skills. As a scout, I took pride in wearing my uniform. I still strive to adhere to the scout law and motto.

As I approached my fourteenth birthday, I would have to prepare to move to another boys' home. Children's Baptist Home housed boys up until age fourteen. One of my best friends, Phillip Mack, had recently moved to a new boys' home in Woodland Hills, California. Phillip had become my best friend at the Children's Baptist Home. He was now housed at the Pacific Lodge Boys' Home in Woodland Hills. Phillip told me about how clean the cottages were there, and how fun and friendly the staff were, and the opportunities to participate in many indoor and outdoor athletic events. Phillip made Pacific Lodge sound like a really fun place, so I told my social worker about my interest and she set up a visit and interview with the intake social worker at the Pacific Lodge Boys' Home. The interview went well, and I was transferred.

So, at age fourteen, I moved hours away from the inner city of Los Angeles to Woodland Hills. Inglewood and Woodland Hills are worlds apart. Woodland Hills is an affluent community with the best high schools in the state. The Pacific Lodge Boys' Home was composed of four cottages and a group home. The rules in the cottages were strict and boys were not allowed to leave campus. Like Children's Baptist Home, Pacific Lodge had one cafeteria where all the meals were eaten. The campus had a huge track and basketball courts. I participated in every intramural sport offered. I competed on the lodge's basketball, track, and swimming teams. Pacific Lodge would compete against other boys' homes.

I attended public schools as usual. As a ninth grader, I attended Parkman Junior High School. Unlike attending public school in Inglewood, the student body at Parkman Junior High School was composed of students from upper middle to upper class families. In fact, the actress Justine Bateman was one of my classmates at Parkman.

I finished ninth grade and moved on to Taft High School in Woodland Hills. I did not know it at the time, but Taft High School was one of the best high schools in the state of California. As a tenth grader at Taft, I was not a very serious student. I literally do not remember studying very much, and my grades reflected it. I tried out for the B football team as a tenth grader and was basically the smallest guy in practice.

I really felt if I made the team the coach would probably have me sitting on the bench all year, so I decided to try out for the gymnastics team. I had some basic talent as a tumbler, tricks I learned watching other kids who were members of a gymnastics club. I tried out for gymnastics and the coach saw my natural talent and he felt I could be a great gymnast. Mr. Gunney was my coach and he would come pick me up from the boys' home. The other team members and I would practice in his back yard, where he had a high bar and a trampoline. I enjoyed going over to Mr. Gunney's house because I could get out of the boys' home. I made the team.

My grades, on the other hand, were mediocre at best. I was naïve about college admissions at this point and I did not realize that ninth grade was an important year as far as college admissions were concerned. So, one day the social worker at the Pacific Lodge Boys' Home called me into her office for a conference since my grades were average. Now keep in mind that California was known at that time for having the best colleges in the country, and without a doubt the best state university system in the country. Graduates of the California State University System were the pioneers of the microprocessor and the programmers for the first routers. The founders of Intel, Oracle, and Cisco, major American computer companies, are alumni of the California State University System.

The social worker, who was a very petite lady, simply wanted to make me aware of what it took to get admitted to one of the campuses of California State University or the University of California System. I believe her name was Vemelah Sing. Again, she was a very petite lady with a thick Indian accent. She basically said she thought I was smarter than my grades revealed. She warned me that if I did not improve my grades, I would not be admitted to any state university in California. Well, given that I was a ward of the court, I knew that on my eighteenth birthday I would have to leave the foster care system and support myself as an adult. I was fourteen, so this represented to me four years to straighten things up, and in reality it was more like three years since colleges really did not look much at your senior year high school grades.

So, at age fourteen, in one of the most populous and diverse states in the country with the best schools in the country, I was shaken by Ms. Sing's speech and decided that very day to work hard to improve my grades and become a better student. My first year in gymnastics went well and we competed against the best teams in southern

California. My eleventh-grade year my grades improved only slightly, and I tried out for the B football team and made it. I started all season and eventually did well enough to be selected as second team all league.

While school was going better, my personal life at the Pacific Lodge Boys' Home was not. The trouble began one day when Phillip and I were in line in the cafeteria and one of the other residents called Phillip a nigger. Phillip immediately punched the guy in the face and the guy fell down. I jumped on top of this guy and we beat him until he passed out. One of our good friends named Marty who was Caucasian then came in and broke the guy's nose with one punch while the guy was on the ground. Well, up to this point, I had been a model resident at the home and Phillip and I were both occupants of the on-campus group home for model boys. The group home allowed us more freedoms in the forms of off-campus passes and less in-house supervision. It took years to work our way from the cottages to the group home.

Well, after the fight the authorities at Pacific Lodge wanted me and Phillip out of the group home, so we were ordered back to the cottages. Moving back to the cottages would be embarrassing and depressing. Instead of moving back to the cottages and being humiliated, I decided to run away from the boys' home as I had done many times over the years. I stayed on the streets for about a week until my great auntie Loraine Porter told me I could come stay with her. Now, Loraine lived in Watts, California, which was literally four hours away from my high school. I moved into her garage where she had a bed and because she had no room in her house since her sister and her sister's husband lived with her. I woke up early every morning and caught the bus from Watts all the way to Woodland Hills to continue my schooling.

My aunt eventually contacted my social worker and I turned myself in. My social worker placed me in a new boys' home in San Pedro, California. I then transferred from Taft High School to San Pedro High School. I was placed in this home the last two months of my eleventh-grade year in high school. I was determined to do well in high school, as I was now sixteen and I knew that I had only two more years until I would be on my own. Where would I go?

My dad had died when I was fourteen years old. He had spent years in prison for killing a man over a card game. I was about three years old when he murdered that man. I did not know him well enough and I can count the number of times we spoke on one hand. He did rehabilitate himself in prison by getting an education and returning to society as a productive citizen. He worked at the Martin Luther King Medical Center for many years. I am not sure what his job title was, exactly.

I attended his funeral when I was fourteen years old. There were many people in attendance. In fact, my mother and sisters also attended. I liked the occasion because it was one of the rare times I was able to get out of the boys' homes and spend time with my real family. Sandra and

Joyce both cried. I, however, did not shed a single tear. I could only think about how he left us in this world along to fight and fend for ourselves. I was not remorseful; I was angry. I wanted answers for what happened. How did we all end up separated by the state? How was our family destroyed? The answers to these questions all went to the grave with my dad because no one else knew the history well enough to tell me. My oldest sister told me what happened to the best of her recollection, but a lot of what went on is missing.

My mother was sitting in the front row of the church and she did not shed a tear, either. My father was very abusive towards her. My mother is a tragic figure; when my father was imprisoned for killing a man, she turned to alcohol. She did in fact become an alcoholic and suffered a stroke in her late twenties. She is paralyzed on one side of her body and requires a wheelchair for traveling. My mother's speech is slurred and her memory is not good. I do still see her on occasion when I visit Los Angeles.

*****

As a teenager I matured quickly. Realizing that I would be on my own in two years, I dove deep into my studies. My stay at a boys' home in San Pedro didn't go very well as the counselors and I constantly argued. At the beginning of my senior year in high school, I was transferred to another group home in the center of Los Angeles. I will not mention the name of this boys' home because I don't have a lot of nice things to say about it. It was a group home that housed about eight boys, and many of them were troubled teens who had used drugs or been involved with gang activities.

My first week in that home, I got into a fight with a boy about two feet taller than me. He was much weaker than I was, so restraining him was not a problem. He apparently had gang connections in the local area; he threatened my life and warned me that I should not sleep in the group home that night. I definitely did not and called my sister to come pick me up and let me stay at her house for the night. She agreed and I returned to the boys' home about a day later. There was only one counselor on duty at all times and they usually did not keep a close watch on the boys. We would meet as a group weekly with a psychologist and discuss our problems. These weekly group sessions, I thought, were a waste of time; however, I participated and put in my two cents.

    I attended Manual Arts High School, which was about twelve blocks away from the boys' home. I had to catch the local bus to get to school every day. Manual Arts High School had about four thousand students when I was a senior there. I was determined to do well in school, as I knew I would have to take care of myself soon. I attended every class and did all my homework. I even took an extra computer class in COBOL at the local junior college. I did well, achieving a better than a 3.9 GPA for my senior year in high school. I was named to a list of A students in the

Los Angeles Times and I was the only student from the inner city to win a district-wide essay contest sponsored by MADD (Mothers Against Drunk Drivers).

I had sent off college applications to three schools: CSU, Chico, CSU, Northridge, and Marymount College. I am not sure how I came about selecting these colleges; however, I knew that to have a fighting chance I would have to attend college and get an education. This was a very critical point in my life. I really didn't have any family I could live with on a long-term basis. I had grown up feeling isolated. It is really hard to describe what a person feels like who is placed in the world and then told their family doesn't really want them. Before I received my acceptance letter to college, I didn't think my chances were that great at getting accepted at a large state university in California. My grades in tenth grade were not that great and my grades in eleventh grade were mediocre. Looking back now on high school, I could have done much better in those classes; however, given the situation of being in an unstable environment, I really didn't focus. I diligently worked throughout my senior year, though. I checked my mailbox daily for any news from the universities.

I do not remember the exact date, but it was early in the second semester of my senior year when I received word from California State. The letter opened with "Congratulations. . ." And I don't remember the rest. I was excited! Even as I am writing now, I can remember the joy that this letter brought me. Acceptance to college for me was the ticket to my future. I knew I would be able to advance myself and secure a job one day to take care of myself. My acceptance to California State University, Chico, lifted a huge burden off my shoulders. I had kept all this anxiety inside about the future beginning at about the age of fourteen. What would my future be like? In fact, there are many children in the foster care system who worry about how they are going to take care of themselves once they are released. Many children I spoke with felt that joining the Army was the way to go. I did not want to join the Army at that point in my life. There would be a time for me to join the Army later in life, and when that opportunity presented itself I would do just that, but not as a senior in high school with dreams of becoming a lawyer.

I was headed to college! I felt very strongly then that things in my life were getting better. My childhood had been quite sad with feelings of hopelessness and frustration, constantly running away from one boys' home to another, running with neither direction nor purpose. As the musician Bob Seger would say, I was "running against the wind." At last I began to feel as though I had a map. My path would be to finish high school, attend CSU, Chico, and become a successful lawyer as I had seen Perry Mason portray on television as a child. Little did I know that my future would present new twists and curves far beyond my imagination, or that my family—particularly my brother—would travel such vastly different paths.

Chapter 2: A letter from Royce Wayne Timmons to Chris Hall, edited for clarity

Hello, people.

I'm housed here at Richard T. Donovan Prison. I was sentenced to 1300 years with two life terms. Cruel, right? I must admit I feel I deserved the time they gave me. It all started when I was working for this company stacking bricks on pallets. I was getting paid $500 a week under the table. It was a young brother who had a spot like labor ready. Matter of fact, that's where he learned how to open up his business, so he paid us under the table because of tax reasons. The story starts off how I met this girl, and she knows who she is, but let's say her name is Betty. My life has been hell ever since. Life has been a battle, boy. So this is for you young men and young boys so you won't get yourselves in trouble. It's about making the right choices in life.

I had odd jobs most of my life. I was the only one who didn't want to work, and I thought hustling was my way to the top, but I never thought using drugs would change me into a ruthless devil. I met this youngster on my job, he was cool but I saw his mama and I knew I could fuck her, so I went to work on the youngster, having lunch with him, giving him slack on stacking bricks. Yes, I became the head man, so I was in charge, but instead of being their boss, I was the cool dude letting them off early for a certain amount of money. We worked from 3:00 pm until 1:00 am. I could have made it good for me and my whole family, but I was a dope head. But I thought by me not using every day I was better than the other dope heads, but let me tell you something. No matter if you used once a month, you are like the others. I was a bad father and husband, period. I don't know how to explain everything, but one day I might write a book and tell you my whole life story, uncut.

I tricked the youngster into letting me come over to his house. I met his mother and she was feeling me, so I would buy her son packs of cigarettes and feed him so he would tell his mama and she was impressed with me. She had a piece of a boyfriend, he was a bum but I was a better bum, so she told me to come over Saturday and play cards. I was cool with that. Saturday, I got up, showered, and put on my Rockwear gear, got in my car, and bounced. I was married, but I was a piece of shit because of the way I treated my loyal and good wife. I'll explain later. I had my paycheck in my pocket and I flossed to the tee. I walked in and I saw the lust in her eyes, so we played cards, smoked some bud, and had some drinks. So she told me old boy was coming over to fix her car. I said I understood, so I would leave. She asked why I had to leave, so I said I can't spend the night. She asked why not. Her kids took her man toy back to the Amtrak station so I could spend the night, and it all started.

I lost my job, so I was going to her house daily. I would get into arguments on purpose with my wife so I could leave, but I made the biggest mistake of my life when I left my wife. I fucked up. I was drinking daily. I would smoke weed. I wasn't looking for any work, but one day her son came home with a job and I wasn't having that shit, so I went to a temp agency and got the same job so everything was good, but her ex-husband was always around and I started thinking a certain way. It was easy for me to fuck her, so I felt she was fucking him and our neighbor. I think that was what took me to the edge. Lesson one: don't mess with a homewrecker and be sure about your life in general. I'm going to tell you how I landed here, somewhat. I want to tell you about this journey.

I got arrested January 25th, 2005 for multiple counts, 27 to be exact. I was through. The police said "I'll see you in 200 years." In places like these it's about survival. If you are a street type of person then you have a chance of survival. The cliques in here are whites with whites, Mexicans with Mexicans, blacks with blacks, and others with others. Now, ever since the so-called water shortage, we shower four days a week, Tuesday, Thursday, Saturday, and Sunday, but here comes the twister. They put timers on all the toilets. When you flush, it takes over a minute. When I had unlimited flushes I would sit down and flush, then pass gas and I would take a dump, drop one flush, and I even sit to piss so no piss can get on the toilet or splash on the walls. When I came in, I wasn't doing any of that stuff. I was lost not knowing what to do. I am 58 years old and on September 21st [2017] I'll be 59. I don't look it or feel it. I'm in prison, but not the mainline. I felt that I wasn't ready for all the politics. You can't hang with any other races and you can't eat after another race. You would get beaten brutally or even killed. It's really a war behind these walls. Even on this side it's still the same. They have a gang they started up called the 2-Fivers and they control certain things. They still have certain people from certain races that are still political. They have stickings and people fight and jump on each other. I've accepted my fate. I've done my crime and now I must pay. I came in cut out of my fucking mind. I even went to Patton Hospital, but I was stoned as I could be. I think the judge felt sorry for me. The physician felt something was wrong with me and gave me a 90 day observation. I went there and got into street mode. They have women there, crazy and not so crazy. I didn't care, I just wanted some pussy. They give you rubbers also so you won't get a disease, but I didn't use them because I just didn't care, all I wanted was sex. I used to give a girl a soda, or pouch of tobacco, a cigarette, or they would just give you some sex. I wasn't trying to learn anything. I was one of those types who had given up on life 90 days exactly. They kicked my ass, and they said I was competent to stand trial. I was in the county jail and I got drunk one day and I saw this Mexican sock this black guy I knew so I ran to help him. BIG mistake. I was fighting one of them then I was waking up in the hospital with my ankle broken. When I got back to the jail some black guy was telling me what happened. I wanted to whip his ass because he should have been fighting instead of looking. I learned then being on this side you're on your own.

    Let me tell you what I've learned. Be humble within. Have that yearning to be a better person. This is not the place for you. Patience is everything. I'm not involved myself with the drama, nor do I involve myself with the riffraff. Some people live for trouble. I'm not one of those people. If you asked me would I do my whole life over differently, that's a big yes, but would still want to have my wife and kids because she was the best thing to ever happen to me. We met while I was in jail. She worked for the L.A. County sheriff. I was working as an ODR in the jail staff kitchen, and I was a player and a big user of women, so I wound up with her but after we got married she said "If I knew you had a second life I would have never married you." But she was one woman who went by her vow and until death do us part and through sickness and health, that's why she never left me. We were married 3-15-84 up until June 2004. That's when I left her for that slut, but when she passed away she was still married to me.

I went to Lancaster State Prison September 3rd, 2009, level-4. I was in a zone because of all the medication I was taking. I told the doctor I used to take thorazine. I just wanted to be left alone and be at peace with myself. At first I knew nothing because I had moved into an out-of-the-way cell and I was content, but I didn't have a lower bunk, so they moved me into a cell with a white Crip and he was involved with it all then. The next thing you know, I was doing white, which is a street name for crystal. I was doing shit that made me crazy—drugs. A lot of people depend on that and it's easy to get drugs in places like this, but drugs always bring drama. I only tried it three times. I just couldn't go down that road again. Being locked up for the rest of your entire life is not good. You are stripped of your dignity, and you will never be complete because you become somebody different because you have to. You eat basically the same things but just on different days. Everybody lies, we get food, salad, pizza, chicken, stuff like that. Everybody tries to be better than the next man because he might have more money, gets visits, it's just like that. I take walks around the prison yards and listen to my music on the mainline. You can't wear headphones because you would be in violation of one of the prison codes. If somebody was getting ready to do something to you, you wouldn't be alert, so you have to be on point. I'm an old man and we get the respect because of our age. Somewhat we get a pass, but if we do some scandalous shit we don't get a pass. We go to breakfast about 6:30. The people who work behind the wall go before regular chow. I'm a diabetic and I have high blood pressure. I'm not all bad, but I don't have a diet. I take shots, I'm type 2. If you treasure your life, never go down the road I've taken for life or anybody like me. We live for pictures of women in short skirts, bikinis, eye candy, we live for that. Now what kind of life would that be for you? Why would you hang around losers? Be somebody for yourself and your parents. I'm writing these words hoping they teach someone and change their lives. I wouldn't wish this on anybody. The COs are here to do their jobs. Some go overboard, but most leave you alone if you stay out of the way and program. I'm a programmer now. I'm not saying I'm a saint but 90% of the time I'm doing the right thing.

I've been down for 12 years and three months. I've been to Lancaster Prison and Tehachapi Prison. This prison is very laid back. It's for high risk medical inmates. I'm diabetic, like I said. The things you can do here, you can't do at other prisons. You will get beat down and told you can't do that. We can bring a bowl and take our food out but other prisons you can't take food out. They have a lot of us that are mentally ill in some kind of form or fashion, like myself. You can get mentally strong if you think positively, which most of us don't. All we do is try to work somebody. What I mean by that is use somebody.

I'm giving you my prison thoughts. I'm confined in my cell for 18 ½ hours a day if we go on lockdown. We're housed 23 ½ or 24 hours if we don't shower. We shower every three days, but here there are no lockdowns. On the lower tier they have a lot of rolling walkers and we have a lot of old men. I wish I could stay here until my maker takes me away. Nobody wants to go to a prison where there is always something jumping off being in the warzone. I have a temper, but I hold it in now because I'm not active anymore, where I would bust you in the head with a canned good or batteries. Sometimes in your life you just know it's time to lick back and allow life to take its course. Inside of me I always say what if my last months on the streets were really good? I had gotten two permanent jobs and that never happened to me in my life. I only had temp jobs, so to achieve that I was on my way to do good, but my weakness for slutty women got the better of me. I picked up a whore and got some crack and went crazy. I did things that I never dreamed of doing, but I didn't know at the time but the devil got into me like never before, so I went on a crime spree. I'll let you find out for yourself why I'm here. It's bad and embarrassing, so don't hate me. My brother wanted me to write about what goes on with my everyday life, but it's really simple. I watch TV and I have shows I watch starting at 8:00 pm. I listen to music. I love older songs and ballads. They soothe my mind. They get me through the day. Basically, you do the same thing every day. I have two people I deal with daily, my cabbie and my road dog. Every day we play cards and walk the track and talk about people like everybody in prison does. I'm very fortunate to be one of the guys that has his own and let me tell you there's nothing like having your own if you don't have anything you've got—treated like a bum, they talk about you, saying you're always begging somebody for something.

    The world has changed in 12 years. I'm lost. If I were to get out, what would I do? That's the big question. I know I wouldn't go to drugs and sleazy women. I would have to get some kind of job. Follow your heart and if you have children put them first in your life. Be there for them always. I was a deadbeat dad, period, but I do love my kids with the way I THINK love is. You get very lonely in places like these. There is no one to warm up to. Now, they do have gay guys here, a lot of them, including the guys who use the others. They are also gay. Just because you don't suck or get fucked and you are the one that's doing the fucking or getting sucked, then you are all a homosexual, period.

    I was in the fast lane on the street decades ago and I enjoyed the glamourous life. I never had a real normal life before. My brain was always altered, so you never think about real life. You feel life is all about getting high, money, and sex and dressing nice. A lot of us who were in the game did a lot of pretending to impress women and anybody else you liked hearing about you around in the street. People talk about you in jail. If they know your business, they will spill it, not knowing they are doing something wrong. I'm not proud of my life because I used people and I lied to people. Yes, even my loved ones.

I'm writing this for my brother. I love him so much. He was very disappointed with me because I got myself in a situation that I can never get out of. Chris, I'm so sorry I left you without a brother. I'm glad you gave me a chance to express myself. Thank you for just being my brother. You became a doctor and I became a prisoner for life. I hope my words help someone to better themselves. The streets are not the place to be. Schools will help you down a good path, being positive and true to yourself. Treat your wife or woman like a lady is supposed to get treated because if you have a daughter now or in the future, you wouldn't want her mistreated. Be honest, period. If somebody slips and falls, help and pick them up. Thank you, Chris!

Mr. Royce Wayne Timmons
G58736-18-224
480 Alta Road

San Diego, CA 92179

Maybe a book in the future.

Hall 52

Chapter 3: College

I moved in with my sister on my eighteenth birthday. This was March 1st, 1984, and I had about three months until I was to leave and attend CSU, Chico. Sandra and I were roommates and I continued attending Manual Arts High School. I was able to purchase a car with some money my dad had left me from his life insurance. I continued to work hard in school. This was really the first time in my life I had real freedom. I was no longer under state custody, eighteen years old, accepted to college, and cruising in a beautiful red five speed sports car. I would drive all over Los Angeles, out to the San Fernando Valley to visit old friends at Taft High School, and out to Venice Beach. Two months after buying this car, all the fun ended.

One night when I was coming home from the Valley and driving back into Los Angeles I pulled over on the side of the road to check my tire. However, about a minute before I was about to exit my car, a large Ford truck struck me from behind and smashed the back of the car all the way up to the driver's seat. Luckily, I wasn't hurt. I didn't even have a scratch on me. This was a horrendous crash and I am lucky to be alive today. If I had exited that car just a minute earlier I would have been dead. My car was spun three hundred and sixty degrees and I was able to free myself from it.

    The young teenagers in the truck that hit me came back to check on me. I informed them that they were in pretty big trouble. The driver turned out to be intoxicated. I went to a doctor and received medical care for six months. My time in Los Angeles was coming to an end and I was really ready to head to college. I wanted a new environment. I had grown up in Los Angeles among the gangs and constant loud sirens. I was ready for a change.

    I attended a summer program through the EOP. The EOP is designed to help underprivileged teens succeed in college. I was happy to be a part of the EOP program that summer and I made some really great friends. Throughout my four years at CSU, Chico, EOP provided me with financial and educational assistance. It was a very supportive program.

The campus at CSU, Chico is beautiful. I attended CSU, Chico back in the early eighties. The campus was full of tall, green leafy trees and fashionable landscapes. There was a beautiful creek, low flowing, that ran throughout the whole campus. I do believe that CSU, Chico has one of the most beautiful campuses in all of California. As a state resident, I only had to pay registration fees. These fees were extremely low, around four hundred dollars a year. Attending Chico was an excellent option—low-cost, high-quality education.

Moving from the inner city of Los Angeles to a small college town was quite a change. In Los Angeles, there were many nationalities and, basically, many different people from all over the world. Chico was predominantly white, and the students were mainly from northern California and came from middle to upper class families. I was one of fifteen thousand students, similar to most—eager, filled with great hope and ambition. I was different from many in that I had grown up without my parents, wasn't from the middle or upper class, and had come from some eight hours away from the crime-ridden streets of Los Angeles. I took twelve units my first semester of college.

I was determined to remain focused and work as hard as possible to do well in college. I had heard so much about how college was supposed to be much more difficult than high school. The classes were about thirty to forty students each. However, some introductory classes had two hundred students in them. Students in college are much more competitive than students in a typical high school classroom. College is usually a stepping stone for most people who are headed to graduate school or directly into a career. I would agree that the workload in college is much heavier than high school. In fact, it seemed as though all of my classes required a ten page term paper. Typically, one would have to select a topic and start writing early in the semester to achieve a solid paper.

I continued my interest in gymnastics during college. I initially just wanted to take a class in gymnastics to help relieve the stress of studying. However, once I spoke with the coach, he was impressed with my skills as a gymnast and wanted me to come and try out for the team. I was initially reluctant. I didn't want anything to interfere with my education. I eventually decided to give it a try, thinking that if it affected my studies I would just quit the team. Training in gymnastics is very intense. We usually trained for three hours every day. Our coach would lecture on a particular technique and then we would practice every step meticulously. My teammates were a great group of guys and we would often tell jokes and practice at the same time.

After being in class all day, it was simply relaxing to go into the gym and train. I made the starting team my freshman year and we competed against some of the best schools in the country. We were technically a Division II school. However, we competed against many Division I schools. We rivaled with Stanford and UC Davis, to name a few. Being on the road was exciting. We would travel to San Jose State or Davis and stay for the weekend. We would have time to visit malls and go out on the town.

I had such a great time as a freshman in college. I lived in the dorm initially for six months. I had to move out because during the holidays the dorms closed and I had nowhere to stay. I moved into an apartment with three other guys and believe it or not, my rent was only one hundred dollars a month. I spent most of my time at school and really only went home to sleep. I studied like a man possessed. I would finish my classes about four in the afternoon and then go eat. I would make it to the library about 5 PM and then study until the library closed, which was usually about 2 AM. Once the library closed, I would go to the local donut shop and stay until four in the morning. My method of studying was very time-consuming. For a given class I would read every chapter assigned. Then, I would summarize every chapter so that when I went back and studied the material for a test I would not have to refer to the book. Lastly, I would do all the problems assigned until I answered every question correctly.

Now this was a very time-consuming way to study and it would not be until I entered graduate school that I would really understand the difference between and passive and active learning. However, my way worked fairly well and I was able to achieve a grade of A or B in mostly every class I took in college. I entered college with an ambition to become a lawyer. I quickly became bored, though, as a political science major. I was quickly learning that what interested me was science. It began with freshman biology and trigonometry. I did well in these classes and this in turn motivated me to take more interesting classes.

I clearly had an interest in math stemming from back in my high school days. I consistently did well in math; however, I did not want to be a straight math major. I did want a major that had math in it. Physics and engineering were too math-oriented. I had to fill my science general education requirements, so I decided to take chemistry. I had not taken any chemistry in high school, so I was interested in finding out what chemistry was about. I chose to take chemistry for majors as opposed to introductory chemistry. I probably should have taken introductory chemistry first; however, I went and looked at the book for chemistry majors and based on that I simply felt I could do it. In the college catalog it mentioned the requirements for the course as two years of high school algebra and one year of high school chemistry. Although I did not take high school chemistry, I decided to read an introductory book over the summer and take it in the fall of my second year.

   The first year of college went well and I achieved very good grades. There I was, once a ward of the court who had grown up in state custody, and now at nineteen years old I had completed my first year of college at a major California state university and had done well. I was proud of myself. I was beaming with joy! It was my little secret. I never told anybody how I had grown up, and so it may have been hard for my peers to understand why I was so happy. In fact, I had kept the promise to myself to do well in school and remain focused.

My first summer in Chico I worked multiple jobs. I worked at Burger King and I worked as a janitor at a bar cleaning up in the mornings. Looking back on it now, I guess I was struggling, but I didn't realize it. I was simply caught up in being independent and mapping out my future. I felt that at that point in my life I could be anything I wanted to be.

I had a fun summer in Chico, going to parties and going out to the clubs downtown. Downtown Chico in the late eighties had popular dance clubs that were frequented by college students. In the summer the town was really quiet and peaceful. The cost of living was cheap and so it was easy for me to take care of myself and stay in Chico.

The summer passed and I was eager to begin the fall semester of my sophomore year. I had done well my first year and so I had gained a bit of confidence. I registered to take chemistry for majors, calculus and physics, and one more non-science class. My chemistry instructor was one of the toughest in the department and students were known for having to repeat his class. I couldn't enroll in another course because my schedule would not allow it. I was trying to complete college in four years, so I had to take some classes all at once.

General chemistry is a conceptually-based class. It is in fact a survey class of all the areas of chemistry. The math in general chemistry is not very difficult. Grasping the concepts, however, is a different story. You study and discuss very small particles that you never see in reality. The problems are not purely mechanical, so going through a step-by-step fashion to solve them will not always give you the right answer. Therefore, having some familiarity with the subject before jumping into college chemistry is a good idea, hence the reason for prerequisites.

Halfway through my first semester of college chemistry I wished I had taken some chemistry in high school. I can still remember my first exam. The room was quiet enough to hear a pin drop. The only sound was the clock ticking away and the buzz of calculators throughout the room. The problems all seemed so difficult. Each tick of the clock made me panic more. I didn't do very well on that first test. I had studied very hard and had literally spent every night studying and studying throughout all my weekends.

I received my test back and it was barely passing. I was shocked and began to panic. I was reminded of my childhood being in the boys' homes and the feeling of emptiness. I thought, "What if I flunked this class and then out of the university?" How would I take care of myself?

This is the resounding theme that keeps coming back to you when you are an orphan. How will I take care of myself?

I took a day off of studying and thought about my life and where I had come from. I had to rebound. Besides, I had two other hard classes to focus on. I eventually focused and decided that I needed to restructure my studying. I would have to do more problems and less reading to do well in chemistry. My grades improved steadily in chemistry and I eventually received a respectable passing grade in the course.

I found the course to be very challenging and decided that if I could conquer chemistry, then I could do almost anything. I declared chemistry as my major beginning the second year of college. It was during this time that I decided becoming a doctor would be a profession with many opportunities. How did I come to this decision? I had proven to myself that I could do well in the sciences.

Growing up in the Los Angeles Unified School District did not afford every student the opportunity to take challenging science courses. What it boils down to is that these classes were limited because the people who taught these courses were in short supply. The students who got to take classes like chemistry and physics in high school were those kids whose parents went up to the school and demanded that their kids be placed in them. Since I lived in boys' homes most of my high school career, I really didn't have anybody to speak up for me and tell the high school counselors what classes might benefit me in the future.

So, here I was in my second year in college academically confident and now wanting to be a doctor. Additionally, I had read a book about a very successful doctor who had grown up poor as a child and worked his way through college and had done well enough to be admitted to medical school. I just felt if another person could do it, then so could I. I continued to work hard throughout my sophomore year. I purposely did not date, and very rarely did I go out to clubs. I knew the competition to get into medical or dental school was tough.

Growing up as a child, I realized that our family was very poor. I had to choose a career that would both hold my interest and bring in some money. Since I had a strong interest in science, medicine or dentistry seemed like the perfect fields. The requirements to become a physician or dentist are very steep, however. Most all medical and dental schools require at least one year of college level biology, two years of college level chemistry, and one year of college level physics. It is not enough to make Cs in these courses, and most medical or dental schools will want you to make all As and Bs, preferably As.

I took my first semester of physics my third semester of college. Again, this was the first time I had taken physics in my life. The lecture hall held two hundred students. Our professor was very energetic and he would randomly point you out in lecture to answer questions. Physics is a subject which is strongly based in mathematics.

I have always enjoyed math, so physics and I got along quite well. It was my interest in math that kept me interested in school. I suggest that young people find an interest and hold onto it and try as hard as possible to develop it.

Even though I was interested in physics, the course was tough. There was a three hour lab that we had to attend each week to test the principles of physics. Throughout the year, I did well in my physics classes and I was happy to complete the series. In my second year of college, I began taking classes during intersession. These classes were usually fast-paced and required a lot of reading. This allowed me to gain more units and assured that I would graduate in four years. I went to summer school between my sophomore and junior years. The beginning of my third year in college was an exciting time. I was a junior in college and, by this point, well known to the faculty in the department of chemistry. This is the year that I took organic chemistry.

Organic chemistry is the chemistry of molecules that make up organ systems. Many of these molecules are large, so their interaction is based on structure. There really is very little math in organic chemistry at the undergraduate level. This class is known for flunking some of the best science students. It was a very time-consuming class, and I pulled an all-nighter for every test in the second semester. I probably did not have to do that. However, my unit load was high and I needed the extra time to do well on my exams. I completed my junior year with good grades and moved on to senior year.

Physical chemistry is the most feared course of senior year. In fact, it was a course that had a reputation of being one of the most difficult courses at the university. Simply to enroll in the class, one must have completed one year of college physics, two years of college chemistry, and three semesters of calculus. It is a math-intensive course. It is a class that explains all of the rules you learned in general chemistry and derives every formula from scratch.

It encompasses thermodynamics, quantum mechanics, and elements of differential equations, to mention a few topics. The class is one year long. It was a challenging class and I did not know it at the time, but one day I would complete a master's degree in the subject.

Really, by my senior year of college, all I wanted to do was get out of Chico. I had enjoyed my time there, but there was little diversity in Chico. I had grown up in Los Angeles around Hispanics, Asians, and whites.

I began applying to medical and dental schools during the beginning of my senior year. I had worked as hard as I could to fill my transcript with As and Bs; however, my GPA was still essentially a B average. I had managed to receive some Cs in high unit courses and this brought my GPA down. I applied to about twelve medical schools and seven dental schools. I was rejected from every medical school except two, and essentially admitted to all the dental schools. I had taken both the Medical College Admission Test (MCAT) and the Dental Admission Test

(DAT). On the DAT I had scored extremely well. My overall science and academic averages on the DAT were both above the ninetieth percentiles nationally. The test covered college-level general and organic chemistry, math through trigonometry, a year of college-level physics and biology, and some aspects of biochemistry. In general chemistry specifically, I ended up scoring above the ninety-ninth percentile. I was shocked when I received my score report.

    I remember traveling across the country on a train to my interviews. It was the cheapest way for me to go. I traveled to Washington, DC to interview at the Howard University School of Medicine. It was a long six day trip. It was snowing at that time of the year and traveling through the snowy mountains of Colorado was refreshing. I caught a cab to get to the school, and I was interviewed by two professors. One was a professor of biochemistry who made me really feel at home. We discussed my interest in medicine and she told me I was a competitive candidate. I had only completed a year of biology, though. I had listed on my medical school application that I would take an upper division biology class in the upcoming spring semester. During my interview, I told my interviewer that I probably would not enroll in the upper level biology class. I don't know if she understood what I told her.

I would find out some time later that she did not understand what I had said. On my way back to California, I stopped in Chicago to interview at the Northwestern University Dental School. Being offered an interview at Northwestern was so exciting. Northwestern was consistently ranked academically on par with the Ivy League schools. The university had one of the largest endowments in the world, and the best and brightest students. If accepted, the boy born in the worst projects in Los Angeles, Nickerson Gardens in Watts, would be on his way to one of the most selective professional schools in the country. They had paired me with a student to stay with. Unfortunately, my money was running low and I was unable to stay for the interview. Further, I had just left DC and I felt confident about my chances at Howard University. I returned to California to finish my final semester of college. I continued to receive invitation letters for interviews from various dental schools; NYU and Boston University invited me for interviews. Due to my finances, I wasn't able to attend those. I waited patiently every day, checking my mailbox for good news. Thin envelopes were most likely bad, so I always wished for thick packages. I unfortunately did receive a number of thin envelopes.

      My first acceptance was to the Boston University School of Dentistry. I was overjoyed! This school was not my first choice, but I was excited because I had turned a corner. Now, one must realize that no one is guaranteed acceptance to any medical or dental school. The competition is generally fierce, and even students with the best credentials are turned away. Additionally, Boston University was considered to be rather prestigious. I waited to respond to their offer, since I wanted to see what would come next.

I soon received my acceptance letter from the Howard University School of Medicine. It was a conditional acceptance requiring me to complete the biology courses that I was enrolled in for the spring semester of my senior year. This was a nightmare! I was not enrolled in these courses, and I had told my interviewer that I would not be taking those classes. I was depressed. I wrote the admissions committee explaining my situation and they changed my acceptance stating that I must take these classes in the summer. I responded to the committee telling them what classes were offered in the summer. They were not happy and told me that I would simply have to reapply the following year once the classes were taken. I could not believe it. There I was, 22 years old, about to realize a dream in one instant, and then having it snatched away in the next. I do not blame the committee, and I appreciated their willingness to work with my situation.

I had been accepted to only one other medical school, but they put me on their alternate list. So at 22, my quest for medical school had ended. About two weeks later I received a letter from the Northwestern University School of Dentistry. It was a thick envelope. I opened it and the letter inside read "Congratulations! You have been selected as one of the best and brightest . . ."

At the time, I didn't really know much about Northwestern. I applied to the school because one of my chemistry professors had done his post-doctoral work there. I told some of my classmates and they were impressed. Many of my professors were also impressed. I had been admitted to one of the most selective schools in the country. I had completed a very difficult major, and although my grades only averaged a B overall, I had scored above the ninetieth percentile on the Dental Admissions Test. I was on my way to Northwestern! It was such an impressive school that I could not say no. I turned down Boston and completed my final semester of college. I was ready to leave Chico; it had been a difficult four years. Chemistry had been tough. I was happy to be done with all those meticulous labs. But I had received an excellent education. I had put myself on firm ground by earning my Bachelor of Science degree. I would always be able to take care of myself.

Chapter 4: Dental School

I arrived in Chicago in August of 1989. It was sunny and the city was bustling as usual. I had applied for and had been given a room in the graduate student housing close to McGraw Medical Center. Once settled, I went and toured the city. I believe I had a week before classes would start, so I went to the park to enjoy the jazz played there on many occasions. I didn't have a car, so I mainly had to ride the L. The L at that time was the major form of city transportation. One could ride from the North Side of Chicago all the way to the South Side. The buildings in Chicago seemed so huge to me at that time. I would walk up and down Michigan Avenue window shopping at the various stores. During my first week in Chicago, all of the first year dental students at Northwestern went on an icebreaker cruise on Lake Michigan. The food was fabulous and I got the chance to meet all of my classmates. They were from private colleges like Brown, Georgetown, and Northwestern with a significant number from the UC schools in California. There were also students from Canada. I was ready to get started with classes and down to business.

I would be in class with some of the best students in the country now. I reflected on my past, thinking about where I had come from. I was a long way from California. Time and space had put me years and miles away from my troubled childhood. However, the truth of my childhood lingered inside my soul. There were still tough questions that would not go away—Why me? Why was our family ripped apart by poverty and crime?

True, I was at a very prestigious dental school; I had accomplished much to be proud of. Why, then, did I still have this feeling of emptiness inside? I thought a lot about my family. We had all endured years of being displaced. Royce and Joyce had grown up in the city of Los Angeles, and they had grown up in a tough part of the city. I had not seen my older brother Royce very much growing up as a child. I really didn't know much about his world. He had been incarcerated at a very young age and up until this day has spent most of his life behind bars. I am not sure if his upbringing had anything to do with his subsequent life as a teenager, or as an adult. I will let him speak on that. I would have to ponder these tough questions later. I had to focus now on my first year of dental school, which was very challenging. The first year classes included anatomy, biochemistry, physiology, histology, neuroscience, and behavior science.

    I was studying around the clock. I would study in the huge library at the Northwestern University School of Law. Medical, dental, and of course law students would use this library to study in. The basic science buildings were next to the McGraw Medical Center. Each floor housed a different department. Anatomy was the most challenging class for most of the people in our class. It involved the gross dissection of the head, neck, abdomen, and back regions. There literally are hundreds of nerves and blood vessels that have to be learned. Not only do you need to know their names, but equally important are the functions and disease outcomes associated with abnormalities which may occur in the vessels and nerves. Anatomy simply required a lot of memory. Our professor explained to us that it would be like learning a new language. I did what I had to in order to get through the course.

While anatomy was not my favorite course, I enjoyed biochemistry. The chemistry of life is simply fascinating. Our bodies are essentially made up of repeating units of large molecules. Biochemistry involves the study of not only how these molecules are arranged, but also what functions they serve. The tests in biochemistry, however, did not focus on the normal function and shape of these molecules. The tests were tricky and would challenge you to think about what happened when normal function and structure were altered.

I made friends with some of the medical and law students and we would periodically go out to dinner. I enjoyed myself in Chicago. I went out much more than I did in college. I made sure I would get at least five hours of sleep every night. In college, I would sometimes sleep only three or four hours a night. I felt I had worked hard enough getting into dental school, and decided it was time to cool down and lead a less stressful life. On the Northwestern campus in Evanston, I would work out at the gym and visit the bookstore.

During the Christmas Break I returned to Chico to visit old friends. Even though I was happy to be out of Chico, it still was the town up to that point in my life where I had lived the longest. Chico had sort of become home for me. Alas, most of my friends had graduated and left Chico.

I returned to Chicago after Christmas Break and kind of abruptly decided that Chicago was not the place for me. I really didn't like the weather. I was from California and was used to sunny weather and skating up and down the beach. Additionally, I yearned for companionship. I had dated numerous girls throughout college, yet I really did not have one girl that I called a girlfriend. I guess I was so busy studying that my social life suffered. I also wanted to meet other young African Americans who were striving to become professionals.

I spoke to the dean at Northwestern and told her about my desire to transfer to another dental school. She was not encouraging and concluded that I would be able to get the best education right there in Chicago. I personally contacted the dean at Meharry Medical College in Nashville, Tennessee, and he said that if I got all my transcripts and letters of recommendation in within three weeks, he would be happy to accept me. I did and so I finished my first year at Northwestern and moved on. I was in my early twenties and still very adventurous. I had read all my life about the South. I knew all about the Civil Rights Movement and the strides Black Americans had made. I never thought much about racism growing up in California. I grew up around many Mexican Americans and Asian Americans, and my high schools were very mixed. I had always attended schools with whites and had made many good friends. However, coming back South was new to me.

I left Chicago in the summer of 1990 and hopped on a Greyhound bus headed south. It was a twelve hour ride and it gave me a lot of time to reflect. I was living my dream of becoming a productive citizen in society. My upbringing had not hindered me, and my future was looking bright.

I was no longer the poor orphan kid from California. I was an educated young man delicately plotting his future. I informed my family I would be moving to Nashville and they seemed rather surprised. I was headed to Meharry Medical College, a college which has trained a large percentage of the African American dentists and physicians in this country, and which is situated in the city of Nashville. It is right across the street from Fisk University. I showed up in Nashville a week before classes started.

Finding a new place in a city foreign to you is always difficult. I located an apartment about three blocks away from the Meharry campus. The school is situated in an underserved area, and most students either lived in college housing or miles away. The student population was diverse, and there were significant numbers of Asian and white students in the medical and dental schools. The class at Meharry was about forty dental students. In general, most of the students were from the South; however, a large number of students were from California. The tuition was sky high—somewhere around $27,000 a year. Most of the students came from middle to upper middle class backgrounds, and many of their parents were also doctors.

I remember being very excited to be in this new environment. There were plenty of young women and I definitely went on my share of dates. The second year of dental school was really technical and involved very little science. I was becoming increasingly aware that this was not the field for me. I was not the only one in the class who did not want to be there. I soon found a number of students who, perhaps like me, wanted to be physicians but somehow ended up in dental school. I personally think the profession of dentistry is a noble profession and highly needed, and we certainly need people who desire to become dentists. However, at that point in my life, I quickly realized that I was in the wrong field. I had completed two years and had taken the National Board of Dental Examiners and had done well. Yet, I was not going to stay in dental school and that was final. I was dating a girl from the medical school at that time and she encouraged me to seek my true goals. She and I became great friends and eventually dated for about five years. Unfortunately, our relationship did not end on great terms. I do wish her the best in all her endeavors, though.

    I decided to approach the dean of the medical school and let him know that I really wanted to be in the medical school and not in the dental school. The dean of the medical school was a very student-friendly man, although very frank and to the point. He advised me that quitting dental school was not a very good idea. He explained that the competition to get into medical school was fierce and that there would be no guarantees that I would be accepted. He explained that every year there were students with excellent credentials who had to be turned away simply because the class could only accommodate eighty students. In short, he was not very encouraging and in fact discouraged me from applying since I was already in dental school. I had made up my mind; no one would discourage me from becoming a physician, even if it was the dean of the medical school, who incidentally was also the head of the admissions committee. Since I had only taken a few hours of biology in college, I had to go back to college and take more biology classes to meet the requirements for medical school. I had taken a year of college biology to get admitted to dental school; however, that had only been seven hours, and most medical schools required eight hours. I promptly withdrew from dental school and began taking science classes at the local university. I worked at night and took classes during the day.

I ended up taking fifteen hours and making all As in the courses. I was also substitute teaching when time permitted and working at night, but was still not making enough money. I began to look for a job in chemistry, since I had a bachelor's degree in the field. I applied for jobs with the State of Tennessee and in the private industry, but could not land a job. The industry for chemists was simply not very large in Nashville, Tennessee.

To my fortune, however, I met a young lady in my cell biology class who knew a physicist at Fisk University who was looking for someone to work in his lab. I went to Fisk and inquired about the position. The lead professor actually was looking for graduate students and he explained to me that if I would come to Fisk as a graduate student it would be easier for him to pay me. The only problem was that my degree was in chemistry and not physics. I had taken the required one year of physics as a chemistry major. One year of physics definitely was not enough background to start a physics graduate program. After some discussion with the head of the chemistry department, it was decided that I would be admitted by the chemistry department and I began taking classes in chemistry and physics. My research would be done in the physics department. I would take upper division physics courses and complete the education curriculum in chemistry.

My advisor was a physical chemist who was on the faculty in the physics department. I will not mention his name here, but he had a huge impact on me. He earned his PhD from Texas Tech University in Lubbock, Texas. He was a true scientist and a hard worker. He would stay in his office until three in the morning and sometimes even slept there. His passion for science was unyielding and he brought in large sums of research dollars from NASA and the Department of Energy to Fisk. We did some very exciting work. My thesis produced three published articles, and I gained a tremendous amount of knowledge from him. He was very hard on me. He wanted me to be a scientist and follow in his footsteps. Truly, my mind was not on a PhD. I wanted to go to medical school. He would tell me I had great potential to be a successful scientist. He wanted to see me at Vanderbilt or Berkeley getting my PhD in physical chemistry.

My time at Fisk was very rewarding. As a graduate student, I worked as a TA and I had to teach general chemistry lab. I tried to make the work as interesting and as easy to learn as possible for the undergraduates. They appreciated my efforts and I began to find teaching rewarding.

I enjoyed the small academic community at Fisk and my chemistry professors pushed me to be the best. They had in fact trained at some of the top institutions in the country (MIT, Yale, etc.), and I absorbed all the knowledge I could from them. I was particularly happy at how things were turning out. Since I had a graduate fellowship at Fisk, I really had no financial worries. I received a stipend every month that covered my room and board. I put all my energy into studying and doing my research. It was a challenging two years under a tough graduate advisor. I did excel, however, and I was able to complete my master's in physical chemistry with perfect grades, achieving a 4.0 GPA.

My last six months at Fisk were grueling. I worked on my thesis around the clock. My advisor wanted my thesis to be perfect. He would meet with me weekly to constantly edit, omit, and even ask me to rewrite numerous pages in order to convey a particular concept. Frankly, the whole ordeal was frustrating and I even thought about taking some time off to figure out if I really wanted a master's degree. I eventually satisfied him and passed my oral defense. I was proud of myself. I had come a long way. I was no longer the anxious kid from the inner city of Los Angeles. I had now achieved my master's degree. I wondered if my father was looking at me from heaven and cheering me on.

It was time to move on—on to medical school. It was a path that I had wanted to take much earlier in my life. Unfortunately, we don't always get to choose what path our lives will take. I had first thought about being a doctor when I was a sophomore in college. I was now approaching 30. I knew this might be my last opportunity to try one more time.

During graduate school I had been meticulously planning to apply to medical school once I finished my master's degree. I had not told anyone. I had been taking undergraduate biology classes at the local state college while I was a graduate student at Fisk. I had completed fifteen hours of biology and biochemistry with perfect grades. I had been studying for the Medical College Admission Test in my spare time. I had scored in the eighty-fifth percentile in chemistry and physics on my MCAT and I had scored at or higher than seventy percent of the test-takers on the biology section. I felt my applications to medical school would be much more competitive than they had been in the past. I had completed all the biology requirements and I had a master's degree this time. Nevertheless, I knew that students with perfect grades and test scores were rejected from medical school all the time. I was focused on this goal and I decided to put my best foot forward. It had been two years since I had talked to the dean at Meharry. His words of discouragement had motivated me to present the strongest possible application.

So, in early 1995, I submitted my application to the Centralized Application Service (AMCAS), which processes nearly all of the medical school applications in the country. I only designated on school: Meharry. Typically, Meharry received more than 5,000 applications a year for a mere 80 seats. I felt confident that my chances for admission were good, and in early May I received a letter inviting me for an interview. I was ecstatic! I had to remember that I had been through this before. I had interviewed at Meharry as a college student and had been placed on the alternate list. I was hoping this would not happen again. I eagerly looked forward to my interview. I read up on the history of Meharry as I had done before and practiced mock interviews.

On the day of my interview I arrived early. As instructed, I arrived at admissions and records and received a list of my two interviewers. To my surprise, one of the interviewers was the dean of the medical school. This was the same person who had discouraged me from applying two years prior. He was my first interviewer, and I met him in his office. He was a biochemist and we talked a lot about chemistry. He was familiar with my undergraduate experiences, so we talked a little about my college. He never once mentioned our previous encounter and he wished me luck in the application process. My second interviewer was a physiologist and he really put me at ease. He praised my MCAT scores and told me he felt I was a strong candidate. He said that he would like to see me in the incoming class and he wished me good luck.

Well, I left the interviews on cloud nine. I anxiously checked my mailbox daily. I hoped and prayed for a thick packet because that might indicate an acceptance letter with other information inside, like loans or scholarships. Two weeks after my interview, I received a thin letter. I opened the letter and read the first paragraph: "Congratulations! You have been selected . . ." My heart raced and I jumped for joy. I settled down and read the remainder of the letter: "as an alternate for the freshman class of 1996." What? An alternate again? I couldn't believe it! Double jeopardy! I slowly sank down in my bed and lay for a number of hours in depression. What were the chances of me coming off the alternate list? I thought over the situation. I had finished my master's with a 4.0 and my test scores were above the school's average for accepted students. Still, I was an alternate. I felt really silly having only applied to one school now. I asked some of my friends who were medical students about my chances and if they knew anything about the alternate list. Surprisingly, many of them had been selected off the waiting list. I learned that it was somewhat standard policy for the school to assemble a long waiting list and then select students from their alternate list.

I did not know what to believe. I stayed in a slump for a couple of days and eventually pulled myself together. I had to focus on finishing my thesis and preparing for an oral defense. My thesis would contain many graphics and I spent many hours on the computer trying to get them right. I had developed a pretty good relationship with one of the assistant professors in our research group and he helped me construct many of the graphics in my thesis. The final product turned out to be a little over fifty pages. I did complete my thesis and I was very proud to learn that some of my work would be published in a reputable science journal. Summer lingered on and about two weeks before the start of classes for medical school I received a call waking me out of my sleep in the early morning. The person speaking asked if I was Christopher Hall and I replied in the affirmative. "This is Mrs. Worthington from the admissions office at Meharry Medical College and I am calling to offer you a place . . ." I honestly do not remember the rest. I was so excited and I simply said yes and she said "We look forward to seeing you on the first day of class." Wow! I was on cloud nine once again. I vividly remember this day and how excited I was. I have to say that it was one of the best days of my life. I had just been accepted to medical school! I was 30 years old and my dream was finally starting to come true.

    I went out that morning smiling with this incredible secret I had not told anybody. I had breakfast out by myself, and I think I went and called my older sister. She was elated for me and was encouraging as usual. It really is hard to explain how one feels once he or she finds out that they will be entering one of the noblest professions in the world. I had always felt that medicine would be perfect for me. I had grown up in poverty, with very little. I was humbled by my childhood experience and had grown into a man wanting to be able to help other people.

Later that evening, I began to tell my friends and they were happy for me. I still had one major problem, and that was how I would pay for medical school. I was no longer eligible for loans since some of my college loans had defaulted. Additionally, getting a scholarship to medical school was nearly impossible since almost everyone accepted had excellent grades and test scores.

I knew that if I applied for loans I would be rejected, so I went to the financial aid office and requested to see the director of financial aid. She was a huge lady and never really smiled. I explained to her that I was going to make arrangements to pay my defaulted loans, and to I asked if she could somehow get my tuition held or waived until I paid at least six months on my defaulted loans. She looked at me and simply asked what I was doing in her office. She said that she really didn't even know why I was there and that if I had defaulted loans that the institution would not be able to give me any money. My head was spinning as she completed her sentence. I remained calm, picked up my jaw, and told her to have a nice day.

I left the financial aid office with renewed determination. It was similar to the determination seen on television when a dying creature struggles to survive. I had come too far to let finances cut my dreams short. I was in between a rock and a hard place, and I had to find something that would give way. I began brainstorming. Initially, far out thoughts came to my mind, like borrowing money from the underworld or charging tuition to a credit card. Then I had more reasonable thoughts, like contracting with a rural city that would pay my tuition and then I would serve their residents once I finished medical school.

Or what about the military? In high school I had not been interested in the military. I was focused on going to college to obtain an education. Ironically, now, twelve years later, I needed the military to finish my education and finally become a doctor. I thought long and hard, and to me it seemed like the military was the best option. The scholarships from the military covered tuition fees and granted the student a monthly stipend to cover room and board. In return, the recipient would serve a year for every year of scholarship with a minimum of three years active duty. Additionally, the scholarship recipient would likely have to complete internship and residency at a military hospital.

    I really did not know very many people who were in or had been in the military. I did think it was great to have the opportunity to become an officer, though. I initially contacted the Airforce and the Navy and told them of my desire. They both instructed me that I had missed the deadline to submit an application. I contacted the Army and they told me the same thing; however, the recruiter told me that sometimes the board would not select all the recipients on its first review, and that if an opening were to become available for this year's class she would contact me. Shortly thereafter, she contacted me and told me that there looked like there might be an opening and to get my application in within nine days.

So, there I was with this 15 or 20 page application asking questions about my life, which dated seven years back. All officers must qualify for a national security clearance, and so the application is very extensive. I thought about my childhood and if any of that information would keep me from obtaining a scholarship. I didn't know the answer, but I knew I had to at least complete the application to have a chance. I completed the application and turned it in promptly. In the meantime, I began preparing for the upcoming year. I was eager to get started and finally set out on the path to becoming a physician.

Chapter 5: Medical School

The very first day of orientation for medical school, all the freshmen were gathered into one auditorium. The freshman class roster was called to confirm that all the freshmen had shown up. The dean of the school of medicine gave us a short welcoming speech. She said that we were a select group (only eighty in number), and that we had been chosen from more than 5,000 applicants. She assured us that we would do well and that the medical school experience would be one of the most challenging endeavors of our lives. I scanned the room to see if I recognized anyone. There were no familiar faces. I learned that the average age of our class was around 26 years old, which meant I was probably older than most of the other students. Our class was composed of students of all ethnic origins and with students from sixty or more different undergraduate institutions. A large number of students were from California. My class was composed of students from

Northwestern, UCLA, University of Penn, Emory, the University of Chicago, and other UC schools, to name a few.

I planned to study hard and keep my face in the books. The first semester of classes consisted of gross anatomy, biochemistry, histology, and behavior science. The second semester consisted of physiology, pharmacology, neuroscience, behavior science, and clinical medicine. The sheer volume of material that must be learned is overwhelming. Learning the subject matter was not conceptually difficult. Medical school classes are not like science classes in college. The classes are more in-depth and much more detailed. The exam questions generally are all in case study format. The most challenging class of the first year is gross anatomy. The class requires the complete dissection of the human body. My school had four students to a body. You essentially start by making an incision in the scalp and then begin cutting. Each skin layer, artery, vein, nerve, and organ is then dissected in delicate fashion. We were tested about every two weeks. The tests were composed of a laboratory component, which lasted about ninety minutes and contained a hundred questions where the student had to identify the structure on the cadaver and then answer some question relating to its clinical significance. This probably was my least favorite class in medical school. I am more of an analytical person. I enjoyed questions that challenged your reasoning ability, not your rote memory. Gross anatomy was rote memory, and you were either good at it or you were not.

    Biochemistry, on the other hand, dealt with mechanisms and questions relating to logic; I enjoyed biochemistry much more. This class assumed you had a basic knowledge of college biochemistry. It was a very challenging class and required intense concentration; it was not unusual for the professor to cover one hundred pages of material in a day. I studied the most for this class and ended up doing quite well. Ten students in the biochemistry class had to repeat it over the summer.

Histology was the third most challenging class of the first semester. This class entailed learning the structure and function of the normal cells that make up our bodies. It really was not a very challenging class. However, it did require a lot of reading and lab time.

In the middle of the first semester I was enjoying my time as a first year medical student when I received a call from the Army medical recruiter informing me that my application had not been selected for a scholarship. She said that I had been a competitive applicant, but that my application had been put in too late. She felt that if I applied early for the upcoming year that I was sure to get a scholarship. I didn't really believe her, but after two months I resubmitted my application for the upcoming fall. In the meantime, I had to borrow money from friends to make ends meet.

It is impossible to work and be a full-time medical student. I was living with a girlfriend at the time since I could not afford rent. She eventually told me I had to move out and I did. In retrospect, my first semester of medical school was very challenging academically, financially, and socially. I could afford top ramen, so that is what I ate.

I was waiting for the Army's response to my application, but in the meantime I had to study and work where I could. Even if the Army granted me a scholarship, I would not be able to receive the money until the fall of my sophomore year. I began tutoring college students at the surrounding colleges to make ends meet. I moved into a small room in a house across from the medical school, which cost only two hundred dollars a month. I borrowed money from friends and family to pay my monthly rent. Since I lived across from the medical school, I was able to walk to class every day. I eventually fell behind on my rent, although my landlord let me stay as I knew her personally—she had been an instructor of mine in the past.

In class, I would overhear some of my classmates talk about how hard things were while receiving checks from their parents and driving BMWs. *If they only knew*, I thought to myself. I did hear from the Army towards the end of my first semester, and the news was good. I had been offered a full-tuition three year academic scholarship. This was significant since the tuition at my medical school was about 26,000 dollars a year, not including room and board. I was elated knowing that I would have the next three years of medical school paid for. About one week later, I was sworn in as a second lieutenant.

Now, I must say that there was a little concern from my family about me being sent overseas to the Middle East or wherever the Army needed me. I never really thought much about this. I thought it was a noble way to serve your country and at the same time take care of America's troops. I kind of went into the military with this gung-ho attitude. The recruiters definitely played up our roles as physicians and officers. We were told that we would receive plenty of respect and responsibility. As students, we would have to serve forty-five days of active duty each year, and this could be done at any military hospital or facility. I was told that at the end of my freshman year I would be sent to officer basic training at Fort Sam Houston in San Antonio, Texas.

Since I would be on active duty for officer basic, I would receive officer's pay, and this would allow me to obtain a large sum of money over the summer. Twelve years earlier at eighteen years old I had not been interested in the military. Now, at age 30 the opportunity had presented itself again and I eagerly accepted the challenge. This time, though, I would be receiving a scholarship worth over one hundred thousand dollars. I could not be happier. The competition to get the scholarship had been fierce and in order to maintain the scholarship on a yearly basis one could not receive a grade below a C in any class, nor could one fail any of the national boards required to obtain the MD. Therefore, I studied long hours in medical school and put my best foot forward.

The second semester of medical school involved taking physiology and histology, as well as an introduction to clinical medicine. This was the first time we would be introduced to clinical medicine. This was the first time we would be introduced to patients and told to do a complete physical and medical history of them. All the science you study in medical school will not help you if you are not able to relate to patients. My second semester went well and I completed my first year of medical school with good grades. The summer following my first year of medical school I was flown to San Antonio, Texas for officer basic training. It was hot and humid in San Antonio when my plane touched down in June.

    I met some of the other medical students who were coming to officer basic. A number of the students were from the military medical school, although some were from elite institutions like Stanford, Harvard, and NYU. My roommate was a student at an osteopathic school and he, like I, had never been in the military. We were not housed on the base, but at a local hotel since the base accommodations were full.

The first morning of orientation for officers was in a huge auditorium on the base at Fort Sam Houston. Approximately one hundred and eighty of us were assembled. We received many welcomes from the important brass on the base; however, the one speech that stands out was delivered by a high ranking military physician who was a general. He basically said that he had had the pleasure to serve for over twenty years and had been in some tight situations overseas, and that he felt that he had impacted a number of young lives and had served his country proudly while doing it. He said that we were all brave and intelligent young people and that we would now be called upon to fill the shoes of two professions, one being a doctor and the other being an officer. I do remember that he stirred up a sense of pride in all of us and we were eager to start our officer training.

The mornings usually began with physical training, starting about five am. We stretched initially and then ran two miles and came back and did more exercises. We would shower, eat breakfast, and then be off to class. I can't discuss exactly what we learned in these classes. I can say that officers must know everything that the enlisted soldiers know. Officers need to know military history and law. They are trained in leadership skills and warfare. The daily classes were long, and we would not get out of them until six in the evening. We did have the evenings to ourselves and most people studied for the weekly quizzes. However, there was time for us to go out on the town and have a good time. Across the street from our quarters was a major mall that had a dance club in it that stayed open until midnight. A number of us would go over to this club and party in the evenings. I made some good friends and we all worked hard to get through our training. I was nowhere near the top of my officer class; however, I did pass and earn my bars. I returned to Meharry that fall eager to begin my second year of medical school. I now had a monthly stipend of about 1,100 dollars which came with my scholarship. My financial worries were now history and I only had to focus on completing medical school.

The second year of science in medical school really is the most relevant year of science as far as clinical medicine is concerned. This is the year that you learn pathology. In the first year of medical school, you basically learn mostly about how the body functions normally. In the second year, you learn a lot about the abnormal anatomy and physiology of the body. As a second year, you learn about the drugs which doctors use to manipulate the physiology of the body. Little by little, you are being immersed in clinical medicine. That is, you are gaining more clinical exposure to patients. The second year by far is the most demanding year in medical school. This is the year that you will have to take your first set of national boards to obtain licensure as a physician. When I was a medical student, the United States Medical Licensing Exam step 1 was a two-day exam. The test was composed of nearly eight hundred questions and your score was dependent on how every other test taker did in the country. USMLE step 1 was a grueling exam. I can still remember the ache in my neck from bending over so long in my seat to answer those questions. This was also the hardest part of the set of three exams required for licensure.

Some classmates and I formed a study group about six months before our class was due to take the exam. We met every night and reviewed the material over the first year and the material currently in the second year. We would quiz each other and do questions from the most common board review books. Our work paid off and we all passed the boards with ease. My medical school would not promote you to the third year unless you passed those boards. Unfortunately, some of my classmates were held back.

For most medical students, third year is the most exciting time of their training. Third year really is the first year that most medical students begin to have maximum exposure to patients. After spending two years studying theoretical medicine, it was refreshing to see live patients being cured of some of the very disease processes you studied in textbooks. Additionally, most students found it much easier to learn hands on. I really thrived in the clinical environment. I made up my mind early on that I was going to try and be as aggressive as possible. I wanted to get my hands wet. I learned how to put in IVs, do pap smears, obtain arterial blood gases, and drain abscesses. Each clinical specialty that we rotated through lasted between one and three months. I particularly enjoyed the junior internal medicine rotation.

      Rotating through those specialties allowed us to get an idea of what field we wanted to go into. I really wanted to choose a specialty that required three years or less of training after medical school. Therefore, internal medicine was the ideal specialty for me. I loaded my senior year schedule with many of the internal medicine subspecialties. I did a month of ICU, nephrology, and cardiology, to name a few.

I also participated in the military match process. The match process in general is a process by which medical students are matched to a particular hospital where they will complete their training for another three to seven years after they finish medical school. The military has its own match process and scholarship students are required to participate in it. The military match began about four months before the civilian match, so military scholarship students knew where they would be training much sooner than other students. Scholarship students must designate at least five military hospitals, and the goal is to get placed at one of your top three choices. I applied to Walter Reed Army Medical Center, William Beaumont Army Medical Center, and Eisenhower Army Medical Center, among several others. I selected Walter Reed as my first choice simply due to name recognition; presidents have been known to receive medical care there. I selected William Beaumont Army Medical Center, located in the city of El Paso, Texas, as my second choice. I had never been to El

Paso, and I knew very little about the city. I would find out plenty about El Paso because when the match results became available, that is exactly where the Army wanted me to go.

It was now about December of my senior year and I knew where I would be training for the next three years. I only had to complete my senior electives and pass the second part of the United States Medical Licensing Exam. I passed the second part of the board in March of my senior year and enjoyed my electives for the rest of the year. I even chose to work with an endocrinologist for a month, where we implanted mice sperm into mice eggs. It was time to move onto internship. I had spent four years in Nashville and I was ready to go. It had been a long road. I made the dean's list in medical school and had overall done fairly well. Like most senior medical students, I was unsure if I was prepared for the next stage of training. I was motivated, and really that was half the battle.

I would be called upon to make life and death decisions. I had heard the horror stories about internship. Our professors in medical school had warned us about the rigorous hours that would be required. The transition from medical student to intern is a steep learning curve. I was ready to move on. The army movers came to my residence in late May and shipped all my belongings to El Paso, Texas. I arrived in El Paso one week prior to the start of my internship in internal medicine.

Chapter 6: Residency

The weather in El Paso, Texas in late June was hot and humid and the temperature was in the low nineties. I landed at the El Paso airport in late June and was taken by a young sergeant to the main entrance of William Beaumont Army Medical Center. The parking lot was packed and the front entrance was busy with civilians and military personnel moving steadily into and out of the building. The medical center was adjacent to the El Paso VA Medical Center and treated all veterans after hours and on the weekends. Additionally, all active duty and military family members were also seen at William Beaumont Army Medical Center. A large portion of the El Paso civilian population also received their care at William Beaumont. It should then be clear why William Beaumont is one of the busiest medical centers in the Army system.

I quickly secured quarters off base about one block from the hospital. This was very convenient, since I had no car at this time. The first two weeks of training we simply went through orientation. Numerous hospital personnel rotated in throughout the day giving lectures on any number of topics from payroll to how to get a clean white coat. My intern class was composed of six doctors. In fact, three of the new doctors had technically been hired through the VA and three of us were sent here directly from the Army. I easily made friends with all the other interns. The medical center in total had about twenty-five new interns in all. The new intern class included interns in medicine, surgery, and a transitional year program. All of the interns had been scholarship recipients except the VA interns, and, therefore, our group was fairly competitive. In general, the new doctor gained his or her experience by doing monthly rotations on different services. Therefore, one month may be spent on internal medicine managing hospital patients or one month may be spent in neurology treating neurological diseases.

The goal of the ideal training program was to slowly allow the young doctor to assume more and more responsibility managing the treatment of patients as the year progressed. Unfortunately, as any physician who has completed their residency can tell you, the majority of programs are not ideal. In reality, the training of physicians can be compared to tossing a newborn into the middle of the Pacific without a life jacket. I was the first intern of the year assigned to take call. That first night was horrendous. My pager appeared to never stop ringing. Not only do interns take call from the nurses on the floor, but also new admits from the ER must be worked up. The calls from the nurses can be about anything. Nurses call about patients that are unresponsive, unruly, bleeding, vomiting, having chest pain, shortness of breath, fever, and numerous other complaints. Sometimes a small discussion on the phone with the nurse can alleviate the problem.

However, in most cases, intervention with drugs or invasive procedures are needed. It is as an intern when new doctors sharpen their invasive procedural skills (lumbar punctures, central IV lines, intubations, etc.), sometimes doing these procedures for the first time and under considerable stress. The nurse would call about critical patients and I would immediately assure the nurse I would be at the bedside in so many minutes.

As a new intern I would have to look up the disease or condition in any number of pocket textbooks, quickly study the material, and be at the bedside in a few short minutes. Residents and interns must be able to study material quickly and be able to apply the knowledge after sometimes only one read. I was called to the bedside of patients bleeding from the rectum, chest pain patients, and lightheaded or altered mental status patients. I cannot list all of the clinical situations for which I was called to evaluate that night. That first night was indeed tough. In the morning after call the intern would have to present all the patients that were admitted to the hospital at night. Most of the time you were sleepy from the night before and you just did the best you could answering questions and presenting the case.

Usually, you had a senior resident over you. This person was usually one year ahead of you in your training. The intern should try to learn as much as he or she can from the senior resident. Some senior residents are very good teachers and will try to teach you everything you need to know to survive in the hospital environment. My senior resident was short on words and on my first day on the ward team had me draining fluid from a patient's thorax. He pushed me, and in the process, helped me develop a tough outer shell. At the time, training under him was misery; however, now I thank him for pushing me to be good.

I think I can clearly say that internship was one of the toughest things I've experienced in my life. Six months into my training things became routine. In reality, after you have seen so many of the same type of patients, working them up becomes second nature. I even spent three months in the ICU as an intern learning how to manage critical patients.

When I had time away from the hospital I toured the city of El Paso. The weather in El Paso was always nice. I mainly spent a lot of my time at the movies. I did not frequent clubs as many of the clubs around the military base were off-limits to soldiers.

I had become engaged to a young lady shortly before I left Nashville. I flew to Nashville and married her four months into my internship and she joined me in El Paso in December. It was nice having her there during such a crazy time in my life. She was very supportive, but I do believe we grew apart as the year wore on. The fact that I was never home may have been a contributing factor in regards to us growing apart. However, being an intern at one of the busiest hospitals in the Army required my full attention. Additionally, I had to focus because I wanted to do my best and learn as much as possible. I did enjoy my time in El Paso, but I could not see myself being there three years. Fortunately, the Army needed physicians to serve in the field as officers with various units.

I thought this might be an interesting experience. It would give me some time away from training and a chance to make a little money. I would be able to complete my commitment with the military and then eventually finish residency as a civilian. I informed the residency director at William Beaumont that I would not be staying to complete the residency and that I wanted to go into the field and become a general medical officer. The director tried to encourage me to stay, but my mind was made up. I informed headquarters of my decision and I was given five possible choices. I would be stationed at a base here in the United States. I was married at this time and my wife really wanted to return to Tennessee. She had been a student at Tennessee State University and she really wanted to finish her final year there. I was in agreement because it would put us back in Nashville, a place that I was extremely familiar with. I received orders for Fort Campbell in early August of 2001. Fort Campbell is situated on the border of Tennessee and Kentucky, and the base is split in half between the two states. Fort Campbell is not a popular destination for enlisted soldiers or officers. It is a high operations base where units are expected to deploy within seventy hours anywhere in the world when ordered. It is a base where physical training is the rule and not the exception. I had about six weeks between the end of internship and the time I was supposed to report to Fort Campbell.

    I remember my final night on call at William Beaumont Army Medical Center. I was the last intern on call of the year. I covered the ward and the ICU. It was a horrendous night. I had the usual floor calls and also the routine admissions through the ER. Additionally, I was called to a code on a dying patient in the ICU. We did bring the patient back multiple times just to have the patient code again and again. I did not sleep that night and when morning came I was happy to check out to the oncoming intern and be on my way home. I was overjoyed to have finished my internship year successfully. I was tired, so all I could do was sleep.

The movers came and moved our belongings to Nashville. My wife had found a small condominium there for us. It was very nice. I must say she did an excellent job of decorating the home. I had never really lived in such a nicely decorated house. My wife returned from a pre-medical summer program and joined me at our new home. I was happy to be back in Nashville. In mid-August I reported to Fort Campbell for orientation and in-processing. The base was nothing like Fort Bliss. The base housing at Fort Campbell was much older. There were no newly constructed gyms, nor were there any shopping malls on the base like at Fort Bliss. The atmosphere at Fort Campbell was not the sunny, slow-paced situation I encountered at Fort Bliss, either. The pace was much faster and soldiers either kept up or were simply left behind.

Physicians in the military have a dual responsibility. They are called upon to be medical advisors to air and land units, which are constantly training for battle. Additionally, physicians are expected to provide healthcare to military members and family members who visit the various hospitals and clinics scattered throughout the military base. Mostly, all doctors are attached to military units and are part of the battalion executive staff and are required to answer to a commanding officer, normally a lieutenant colonel. I was initially assigned to a helicopter unit. This was a new adventure for me, and indeed a learning experience. I had never ridden on a helicopter in my life up to that point. Helicopter units are composed of pilots, crewmen who work on the helicopter, and supporting administrative staff. My job entailed taking care of all the medical needs of the battalion and determining which pilots were medically clear to fly and which ones were not. I was also charged with determining which soldiers were fit for duty and deployment and which ones should be held back. I would also have to put in a full day working in one of the medical clinics on base. It was a very busy and stressful job, and ultimately I had to move closer to base. Nashville was an hour away and I could not commute every day. Doctors had to live within thirty minutes of the base in case they had to respond to emergencies. I found a room in a house near the base and my wife continued to live in the condominium since she was a student at the local state university in Nashville. The separation began to put a strain on our relationship.

My typical day involved getting up at four am and driving from Nashville to Fort Campbell. I would go by the battalion station and treat soldiers there until seven am. I would then report to the clinic at seven am and treat soldiers until three pm, when the clinic would close. At about 3:30 I would stop by the battalion commander's office to see if there were any problems he wanted me to look into. This might entail preparing talks for the battalion or medically evaluating specific soldiers for harsh training conditions. I would also attend executive staff meetings in the battalion and advise the commander on possible casualties that might occur for a particular operation. I really was very busy and I didn't have a whole lot of time to spend at home. This situation was clearly a stress on my marriage and before long we found ourselves in divorce court. We divorced in 2004, although we had not been living together for over a year.

I did enjoy my time in the Army, however. Being an officer and a physician was great. I was able to see three hundred or more patients in a month sometimes. I was discharged from the Army in 2003. I was eager to return to training and complete my residency.

I initially thought about completing my residency in internal medicine. Internal medicine came easily to me and didn't involve treating obstetric patients or children. Finding a second year spot in internal medicine was not easy. The programs that were interested in me were in Boston and New York, and I did not want to go to either of those cities. I began looking at other specialties that were close to internal medicine; family medicine was the closest.

Training in family medicine would give me additional training in pediatrics, obstetrics, and surgery. Additionally, family physicians obtained additional training in emergency medicine since the residency programs required family residents to assume care of patients as soon as they presented to the emergency room or obstetrics screening room. I found a program at the University of South Alabama in Mobile that would give me credit towards the first year that I had already completed in internal medicine. While family and friends wanted me to return to Los Angeles, I actually enjoyed living and training in the South. My sister was specifically concerned about me training so deep in the South and the racism I might experience.

In fact, to the contrary, it appeared to me that blacks and whites interacted more freely in the South than in other parts of the country. I showed up in Mobile three days before the start of my family medicine residency. It was in the middle of summer, so the weather was rather hot. I eventually found a place to live about three blocks from the University of South Alabama Medical Center. Mobile really is a clean city; there are lots of trees and the city is bordered by a huge bay.

I was eager to get started in my training and jumped in with a rather gung-ho attitude. My training was nothing but rewarding. I was able to help a tremendous number of people. The patient population that the University of South Alabama serves is largely underserved. I would estimate that seventy percent of the population is covered by Medicaid.

Family medicine covers a very broad spectrum of treatment. One moment I may be in labor and delivery bringing a new life into the world, and the next moment I may find myself in the medical intensive care unit placing a central line catheter in a dying patient. Since I had completed one year of internal medicine, the director of the family medicine program granted me six months credit.

I needed those first six months as an intern again to brush up on my clinical skills and relearn some of the specialties that I had not seen or practiced since medical school. My first six months went very well and I used my prior knowledge of clinical medicine to excel as an intern. I was promoted to second year at the end of my first six months in January.

Second year involved taking on more responsibility for patient care. You were expected to manage all the patients on the service. Since you were on call by yourself in the second year, you had to master procedures since at any time a patient might require a lumbar puncture or a paracentesis, or simply an incision and drainage of a large infected mass. The service was relatively busy and we had to cover two hospitals. One hospital was simply dedicated to the care of women and children. The other hospital was dedicated to the care of adults. These hospitals were ten minutes apart. One could literally be performing a lumbar puncture on an infant and at the same time be paged to the medical center for a heart attack in progress. One would have to manage time properly to be successful. I enjoyed the challenge and worked very hard. The environment at both hospitals was perfect for a young doctor in training who wanted to learn as much as possible. I had the opportunity to learn multiple procedures—so many I cannot list all that I was able to perfect during my training in Mobile.

    I have about six months of training left as I write this book. I feel extremely fortunate to have been given the opportunity to become a doctor. My life has prepared me well for this task. Growing up in foster care and boys' homes was a difficult time. I remember those early years of insecurity, wondering how I would take care of myself and why my life would be like. I simply had a desire to want more. Although I didn't initially focus in school, I used academics as a way to achieve my dreams.

Having a goal is essential. Once we have a goal, we can then begin to map out where it is we would like to go in our lives. Believing in oneself and one's potential is also very important. I was clearly an underdog and the statistics were stacked against me. My brother Wayne's life, as you have read, clearly took a different path. The choices one makes in life are critical and can determine your future for the rest of your life.

I hope my story has been somewhat of an inspiration to those who feel hopeless and left out. My brother is facing life behind bars for the choices that he has made. I love him dearly and I look forward to seeing him soon. It is five in the morning and I have to be at work at eight. Good night for now, and I am wishing you the best regardless of what life may present to you.

www.ingramcontent.com/pod-product-compliance
Lightning Source LLC
Chambersburg PA
CBHW020433220526
45464CB00002B/680